Massive seracs have formed on the Hotlum Glacier of Mount Shasta.

During the summer, water melts on the surface of the North Palisade Glacier and then refreezes overnight.

The icefall of the Hotlum Glacier reflects warm morning light on the flanks of Mount Shasta.

The morning's first rays of sun strike the Darwin Glacier, seen from Mount Lamarck, west of Bishop.

The Trinity Alps Glacier lingers in the upper left, beneath Thompson Peak—the highest summit in northwestern California. The glacier and snowfields melt into Grizzly Creek, which begins its journey to the Trinity River with this waterfall.

CALIFORNIA GLACIERS

The Mount Gilbert Glacier lies on the left and the Mount Thompson Glacier on the right, at the headwaters of the South Fork Bishop Creek in the southern Sierra Nevada.

CALIFORNIA GLACIERS

PHOTOGRAPHS AND TEXT BY TIM PALMER

HEYDAY, BERKELEY, CALIFORNIA

SIERRA COLLEGE PRESS, ROCKLIN, CALIFORNIA

To Marc Taylor,
His kindness, dedication, and leadership have shown the way to me, and to many.

———————————————————————————————————————

This Sierra College Press book was published by Heyday and Sierra College.

©2012 by Tim Palmer

Library of Congress Cataloging-in-Publication Data

Palmer, Tim, 1948-
 California glaciers / photographs and text by Tim Palmer.
    p. cm.
 Includes bibliographical references and index.
 ISBN 978-1-59714-174-1 (hardcover : alk. paper)
 1. Glaciers--California. 2. Glaciers--California--Pictorial works. 3. Glaciers--Environmental aspects--California. 4. California--Environmental conditions. I. Title.
 GB2425.C2P35 2012
 551.31'209794--dc23
                            2011029758

Cover art and facing page: Blue ice appears deep in a crevasse on the east side of Mount Shasta
Book design: Lorraine Rath

Orders, inquiries, and correspondence should be addressed to:
Heyday
P.O. Box 9145, Berkeley, CA 94709
(510) 549-3564, Fax (510) 549-1889
www.heydaybooks.com

Printed in China by Everbest Printing Co. through Four Colour Imports, Ltd.,
Louisville, Kentucky

10 9 8 7 6 5 4 3 2 1

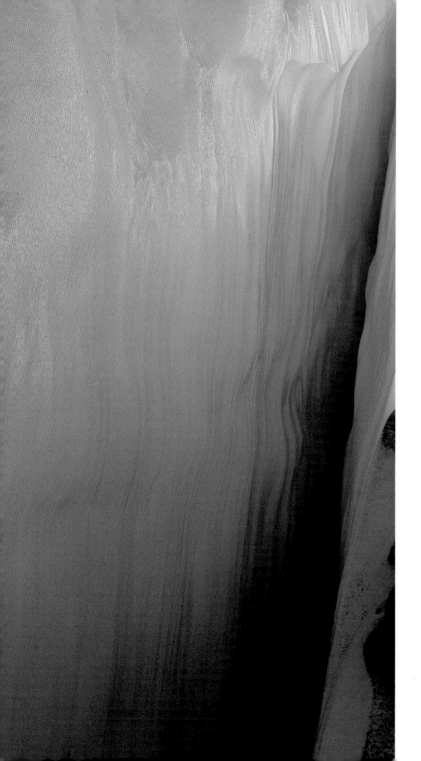

CONTENTS

INTRODUCTION

In the spring of 2010 I set out for the glaciers of California. I wanted to see them before they were gone.

I wanted to climb on the glaciers' steep slopes, to feel the crunch of their snow underfoot, to drink from crystalline streams cutting their icy surfaces, to sleep at their rocky windswept edges, and to photograph their evanescent beauty so others might also know what was there. It proved to be one of the most remarkable summers of my life.

Many people are unaware that glaciers even exist in California. That's no surprise, considering that these small gems of frozen water lie at extremely high elevations, cling to frighteningly slanted slopes, nestle into chilled northern recesses of our wildest haunts far beyond roads and trails, and reside where the weather constantly threatens mortal comfort if not survival. But they're there. They're part of California. They're important, and they're disappearing as the climate changes.

Reaching them was a challenge, but during my spring-through-autumn quest, I visited twenty or so in their high mountain amphitheaters. I was irrepressibly drawn to scenes of lovely elegance and to others of harshest grandeur. I watched nature's workings in phenomena that I never expect to witness again, and also in Earth's everyday chores—one moment, the frightening crush of a five-foot boulder from high cliff to ice surface with a thunderous thud; the next moment the ongoing drip of an icicle, a common yet revealing transformation of water from one form to another.

The glaciers are brilliant: white with drifts of snow, blue with unknown depths of ice, golden with the first light of sunrise, glistening with glassy refrozen veneers at the end of each day, and nearly black but vaguely gleaming and mysterious in their creakings at night beneath the stars. Glaciers and their greater context—the earth, water, life, sky, and

Facing page: Matterhorn Peak rises sharply over its glacier in the Hoover Wilderness west of Bridgeport.

Melting snow becomes deeply grooved and patterned in July at the lower end of the Mount Dade Glacier at the headwaters of Rock Creek, south of Tom's Place and Mammoth Lakes.

weather around them—caught my eyes and propelled me onward with my camera in hand.

Beyond beauty, glaciers provide vital water supplies as they melt in late summer, when mountain streams need the flow. Perhaps most important, the glaciers send a message that we've changed the world in ways that we couldn't have imagined just a few decades ago. Because carbon dioxide is far more prevalent in the atmosphere than it used to be, the climate is warming. The carbon buildup is principally caused by our burning of fossil fuels and by deforestation. The warming results in hotter summers, reduced snow, increased rain, hurricanes and other violent storms, more frequent flooding, intensified droughts, scorching wildfires, epidemics of disease once limited to warmer zones, and the extinction or wholesale decline of plant and animal species that are either unable to move on or unsuccessful in finding new habitat.

Of all these ominous changes, the melting of the glaciers is the most immediately visible. However tragic or trivial one might regard the demise of any single patch of ice, there is no doubt that its melting signifies greater and more threatening changes to come. To add a bit of twisted irony to the old metaphor, the glaciers symbolize just the tip of the iceberg; their melting presages massive changes lurking unseen. For the same reasons that the glaciers are shrinking, the far more widespread Sierra snowpack will diminish markedly. While one might argue that the glaciers don't hold enormous amounts of water, the snowpack does, and its decrease will have profound and inalterable consequences for water supply, nature, and life in California. As the glaciers wane, habitat is likewise running out for plants and animals that depend on the current climate; an unprecedented and degraded world emerges at a revolutionary rather than evolutionary pace. Those who doubt that these problems will grow need look no further than the disappearing ice for a change of mind.

Even though I was burdened by this knowledge, I proceeded with gusto in my once-in-a-lifetime opportunity, inspired by other glacier enthusiasts who had broken trail for me, beginning with John Muir. America's pioneering environmentalist and advocate for wild landscapes journeyed to the California mountains in the late 1800s and prowled around the peaks where the glaciers lay. He later wrote an illustrious series of articles and books aimed at awakening Americans to the beauty and importance of nature, and his first story focused on glaciers. It described how the ice long ago had shaped Yosemite by carving out the valley floor. In another article, Muir was the first to announce that active glaciers still existed in the High Sierra.

Now, more than a century later, I had the privilege of following in that percipient character's footsteps along the trails, across the meadows and moraines, and onto the ice. My task as a photographer was one he might have relished had he been given the tools in my hands, and with them, the opportunity to show others a beauty unlike any other. But my obligation as a writer differs greatly from what

Blocks of drifted snow calve as icebergs into a lake below the Maclure Glacier in Yosemite National Park.

Muir faced. While he joyously announced the presence of "living glaciers" to an unknowing but interested world, my fate is to document and report on these extraordinary features in their final days, or years, or decades, and to bear bad news that nobody wants to hear.

My summer among the glaciers was a time unlike any other in my life—different in an unrepeatable and poignant way, at once melancholic and thrilling. My unexpected attachment to ice led to both despair and to hope, to resignation over something too tragic and enormous to clearly comprehend or even define, but also to a determination that persists beyond the paralyzing path of despair or cynicism. Let me show you what I have seen, and let me tell you what I have learned.

Facing page: Metamorphic rock of the Clyde Minaret rises vertically from the edge of a Ritter Range glacier west of Mammoth Lakes.

With fresh snow in late May, a full moon sets over Mount Winchell (rear, left of center), and the far northern extent of the North Palisade Glacier, which lies in the cirque to the left of Winchell.

CHAPTER ONE

ONTO THE ICE

Though I had planned on photographing the glaciers from all angles, I had never considered being in them.

Yet there I was.

Glacier surrounded me. The sunlit brilliance from the outside world faded behind me as the dark netherworld of ice and buried rock drew me inward.

Imagine a whole mountainside covered in snow: snow ramping up the north slope of 12,960-foot Mount Maclure for hundreds of yards, toward the highest ridgelines and also spanning the full width of the peak's upper basin in Yosemite National Park. The downhill terminus of that glacial mass ended abruptly in a frozen, head-high wall—an edge marked by crumbling ice cubes, dripping rivulets of barely liquid water, rococo sculptures reformed every night by the freezing of new watery masterpieces, slushy mud, and chunks of igneous rock poised to tip or tumble as their foundations liquefied in afternoon sun. That final wall of the glacier had a gap at the bottom that invited me to duck low, scramble forward across a threshold of gritty meltwater, and enter a world unknown but tantalizing in the strange and edgy nature of it all.

The space that I crawled into beneath the ice was created by warmth seeping out of the earth, and by water—warmer than the ice above—trickling across the glacier's floor. Headroom shrank as I crawled farther inside, and when the passage became too tight, I squirmed around to look back the way I had come, and to take a picture. The opening to the outside world was just a patch of precious light filtered through frozen darkness all around me.

I would never have thought to compare this dripping and claustrophobic slice of raw geophysical phenomena to the tidier and more capacious Sistine Chapel, but like the chapel, its most remarkable aspect was the ceiling. Smooth as glass, glossy in gray-black, the surface subtly shined with the tightly rationed light and invited my bare hand to swab its slick surface—like a wet mirror where I sat.

I tried to imagine the weight of the glacier above me and to picture its size, its variable thickness and thinness,

its conglomeration of ice blended with rock and soil like ice cream enriched with nuts and chunks of chocolate, its crevasses and undulations climbing far up the mountain to a bergschrund. That's the gap left when a glacier, as a somewhat cohesive unit, slides downhill in summer from the upper rock wall of the mountain to which it was welded tight all winter and spring.

Glaciers move.

But for now, the icy underworld where I rested seemed stable, serviceably intact, without the rubble of recent cave-ins, rockfalls, or washouts. I felt secure.

Open to it all, I sat, and thought, and imagined.

John Muir would have loved it in there.

Still a young man, Muir hiked to this very same spot in 1872. The idea that glaciers had shaped landscapes during the ice ages had been introduced to the world only thirty-five years before, by the famed earth scientist Louis Agassiz, who theorized that great moving bodies of ice had carved much of the terrain we now know in the Alps and other ranges. Active glaciers were not thought to exist at all in the United States until 1870, when Clarence King of the California Geologic Survey climbed Mount Shasta and became the first reporter to document their presence.

Muir was keenly aware of scientific thought in his time, and with it he combined an enviable wealth of field knowledge and on-the-ground savvy that still makes my head spin. Insatiably yearning to learn more about the land around him, he roamed the Sierra Nevada on foot for days and weeks with little but a bedroll and loaf or two of bread.

It was Muir who first reported the presence of surviving glaciers in the Sierra. In October 1871 he had set out, innocently enough, "enjoying the charm that every explorer feels in Nature's untrodden wilderness." He hiked to the Merced River headwaters, between Black and Red Mountains (Black Mountain is now named Merced Peak), and after climbing above lakes, meadows, and moraines, he crossed a stream and recognized its silty current as "the mud worn from a grindstone." He "at once suspected its glacial origin." The prospects of discovery propelled the thirty-three-year-old adventurer onward. He climbed onto what he named the Black Mountain Glacier, exuberantly strolled the ice, and stopped to peer into the tempting depths of a crack that tapered from a wide opening down to a seam of blue-black darkness. He couldn't resist.

In *The Mountains of California* Muir wrote, "A series of rugged zigzags enabled me to make my way down into the weird underworld of the crevasse. Its chambered hollows were hung with a multitude of clustered icicles, amid which pale, subdued light pulsed and shimmered with indescribable loveliness. Water dripped and tinkled overhead, and from far below came strange, solemn murmurings from currents that were feeling their way through veins and fissures in the dark. The chambers of a glacier are perfectly enchanting, notwithstanding one feels out of place in their frosty beauty. I was soon cold in

Facing page: Sculptural forms of a bergschrund rise in the interface between rock and ice at the top of the Maclure Glacier in Yosemite National Park.

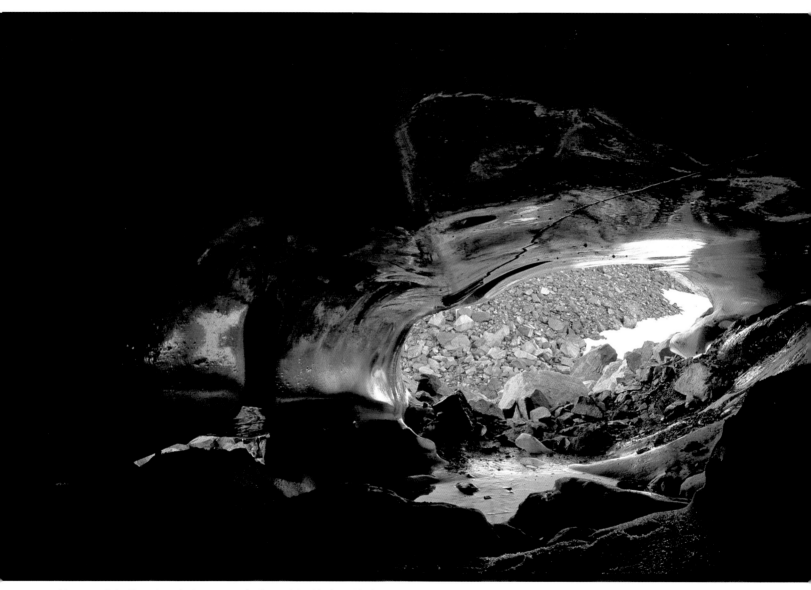

Afternoon light filters into the ice cave at the base of the Maclure Glacier.

Facing page: Elaborate ice sculptures form at the bottom edge of the Maclure Glacier.

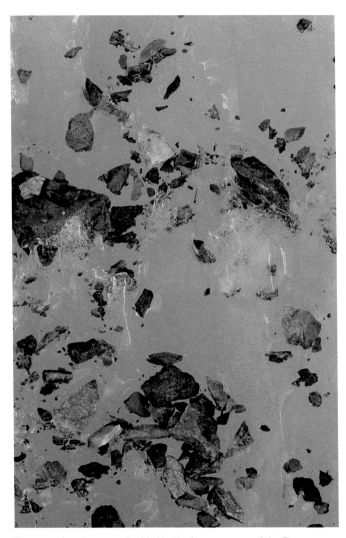

Stones and rocks are embedded in the frozen veneer of the Dana Glacier.

my shirt-sleeves, and the leaning wall threatened to engulf me; yet it was hard to leave the delicious music of the water and the lovely light."

The next summer, right here on the Maclure Glacier, at the headwaters of the Tuolumne River in northern Yosemite, the analytical Muir pounded wooden stakes into the snow in a straight line across the width of the ice and, relative to the rocks at either side, he measured the seasonal downward creep of the glacier. It moved an inch per day, on average, at its center.

Combining his zesty drive to experience the mountains deep in his heart and bones with a scientific bent that challenged experts of the day, Muir was passionate about the glaciers. "You will not find in me one unglacial thought," he wrote to his friend and mentor Jeanne Carr in 1871.

To stir public interest in the places he loved, Muir turned to writing and named his first published article "Yosemite Glaciers." It appeared in the *New York Daily Tribune* in 1871. Later, "Living Glaciers of California," published in *Overland Monthly,* recounted his Black Mountain/ Merced Peak discovery. In response to his theory that glaciers shaped Yosemite Valley, Muir endured ridicule from some of the top geologists of his day, who thought that the shape of the valley owed more to seismic subsidence. But sixty years later, the esteemed François Matthes of the US Geological Survey confirmed Muir's testament in his classic *Geologic History of Yosemite Valley.*

Knowing there was much more to see, I put aside my

ice-cave reverie and thoughts of Muir and crawled out through the mud. I happily squinted into the brilliant sunlight. A few precious hours of daylight remained, so I set out to roam the slopes of the glacier and go as high as I could.

The surface at lower elevations had hardened into ice, though without smooth or glassy refinement; it had no resemblance whatsoever to the pond ice I skated on as a kid. Rather, it was rough, textured, and gray with embedded air or grit, and seemed to have been splattered down by the random bucketful. Rocks, gravel, and sand were embedded in the freeze. All this debris had fallen onto the snowy surface of the glacier from adjacent cliffs and mountainsides and periodically been covered with meltwater that froze overtop, like shellac, so that all these sinkable solids appeared to be floating on water. The gritty cargo rode the slow, downbound train of creeping ice for dozens of years, if not a century or more, while the snow around it melted and froze, melted and froze, day after day, year after year, until the rocks from the summits and ridges arrived down here, at the glacier's end.

The "floating" rocks were evidence that I had entered a totally different realm, and I was only beginning to understand its nature.

For one thing, I couldn't walk in my boots anymore. Though roughly textured, the ice surface was hard, and my waffle-treads didn't grip at all. So I strapped on my crampons—steel hardware that transformed my soles into a bristling set of pointed cleats sharp enough to really mean business. Wearing them, you want to make sure you step on the ice and not on your other boot. Moments before, I had been slipping with every step; now I was dangerously deluded, thinking that nothing would hold me back.

I had long hiked or skied on snow slopes steeper than this, larger than this, and wintrier than this, on both sunny days and in opaque blizzards when a compass was my only ticket home. But even when the snow depth reached twelve feet in the Sierra backcountry, I had always known that it was just snow on dirt. I had dug into the crust to examine the layers of crystals put down by successive storms—layers welded together, or not, by California's signature winter pattern of frequent freeze-and-thaw. The exposed layers revealed, among other things, whether an avalanche was likely, an event I was highly motivated to avoid. Underneath it all, if I kept digging, I knew I'd find soil, and in it, the mountains' tenacious community of life, ready to germinate and sprout into the delicious green of springtime. But here, on the glacier, it was totally different. Ice, not snow, lay beneath my feet. Even a determined swing of my axe would do little but chip a sharp arrowhead of frozen water into the air; it was more like taking a real axe to the trunk of a lodgepole pine than shoveling into a drift of snow. I knew that under the ice lay not soil teeming with microbes, but rather hard rock that had been denied sunlight for centuries. I also knew that the ice moved—or at least it used to move—downhill as gravity bent the solidified water under its own surprisingly pliable weight.

Pit-and-pinnacle formations three feet high form as late-season extensions of sun cups on the upper slopes of the Lyell Glacier.

Here was another glacial truth that required me to stretch my concept of what is real, or likely, or possible: a solid substance that flows. I imagined the creeping ice as if it were extremely dense clay, ever so slowly yielding, bending, and moving to the molding pressure of my hands. Gravity had exerted that kind of force on the ice nonstop for ages, and so the ice bent and slid downhill.

By setting foot on the glacier, I had figuratively—and quite imaginably—reentered the ice ages of the past. I now stood on the same elements that during the Pleistocene had shaped the land, both here near the mountaintops and below, the whole way through the Tuolumne Canyon and the whole way to Yosemite Valley, just the way Muir described all that beauty unfolding long before anybody had arrived on the scene.

As I gained elevation the surface of the hard ice and its mosaic of floating rocks changed; chunky frozen slush covered it, and higher up it gained a veneer with the texture of snow cones—snow halfway transformed to ice with a crusty epidermis not yet part of the glacier but on the way. In some years the snow from the previous winter completely melts, exposing "firn," or incipient ice, all across the glaciers' surfaces. But the winter of 2010 had been stormy and cold, especially late in the season, and so

Rocks and snow cover the surface of the Middle Palisade Glacier at sunrise.

at middle and upper elevations the glacial ice was still patterned by deep drifts of snow that had fallen and blown into place as recently as June.

That snow, however, did not resemble anything we might see on a weekend skiing at Squaw Valley or Alpine Meadows. By early summer, the depth of soft snow that once smoothly blanketed the glacier's upper elevations had been transformed into sun cups. A function of the physics of melting, these depressions occurred all over the snow's surface, creating a texture like sunken teacups or open empty egg cartons. The cups form because the surface of the snow has slight and inevitable irregularities,

Late in summer, the bergschrund at the top of the Middle Palisade has caved in and filled a deep gulf usually left between the glacier and its mountain headwall.

owing perhaps to its deposition by the erratic element of wind, or to subtle differences between gravity's grip on dense snowflakes and fluffier ones, or to raindrops hitting the snow. Even an infinitesimal dent in the surface angles differently to the sun, and so one side of the dent catches more light and heat than the other. The warmed side melts faster. Then, the water trickling off its surface creates a micro-drainage pattern, with newly formed droplets kidnapping additional snow crystals and turning them, too, into water as they seep toward the miniature sinkhole at the bottom of each cup. This erosion of wetted snow accentuates the dented effect.

The extreme of this phenomenon, full-blown in September when I visited the Maclure and Lyell Glaciers, went beyond any resemblance to teacups or egg cartons. Rather, the depressions had deepened to pits dropping one, two, and three feet with steep vertical or overhanging sides. Furthermore, the boundaries between the pits had melted together, leaving razor-sharp ridgelines with peaks angled knifelike to the sun. Walking across fields of this peak-and-pit topography required stepping from one sharp ridgeline to another, with any miscalculated stride fully capable of spraining an ankle.

I walked slowly.

Even so, step after step took me upward with steady progress, which I weighed against the coming of night. I concluded that I could make it back down to my tent,

Dripping water has frozen on cliffs at the headwall of the Maclure Glacier.

At a moraine beneath the glaciers of Mounts Dade (left) and Abbot (right), I found a nominal tent site.

provided nothing went wrong. The slope steepened near the top, and the ice grew harder with the elevation and with the chill of nearly perpetual shade cast by the looming bulk of Mount Maclure's summit, so I stomped the steel points of my crampons down with vigor to gain traction. Kernels of ice rattled into the pits beneath me with every step. A crevasse where the ice had cracked open and exposed the innards of the glacier lacked the convenient zigzag ramps that Muir had enjoyed, and I gave it wide clearance. I would have been concerned about falling unawares into such an abyss earlier in summer, when snow still concealed the cavitied gaps, but now, in September, the snow had melted or metamorphosed toward ice, and the openings were clearly evident. Or so I trusted.

At the top of the glacier the view was breathtaking. Mount Maclure rose as a monument of rocky tablets another hour's climb above and to the west. Directly below the glacier and below the rock rubble of its moraine, a lake curved like a gorgeous and reflective lens on granite bedrock. Flowing from it, Maclure Creek carried the bubbling runoff of the glacier. It ran with a refreshing robustness this late in the season, when all streams lacking glaciers at their sources had faded to mere trickles. The stream disappeared over a convex slope that from my perspective dropped steeply into a void, but I could see the Lyell Canyon of the Tuolumne River farther down, and farther yet, I pictured the epic descent of that queen of Sierra Nevada rivers through its Grand Canyon and on to its reservoirs and lower meanders, all nourished by runoff from the glacier where I stood. Uncounted peaks like granite incisors and volcanic molars rose along the Sierra crest to the north, including the Pleistocene-sculpted faces of Gibbs, Dana, Conness, Matterhorn, and dozens more jutting from the spine of America's greatest mountain range.

I took a deep breath of satisfaction: I had arrived at the top of a glacier. And a breath of anticipation: there was much more to see.

Just above me the bergschrund gaped open. I yearned to scramble into its depth—rock on one side, ice on the other—but for that adventure I'd wait for an early morning, when the snow, ice, and rocks overhead would be frozen tighter and less likely to fall.

There at the top, I sat on my pack in the snow and let my eyes swim in the scene below, relaxed in deep appreciation that my life had, at that moment, intersected so intimately with the life of this landscape. With my eyes following the lines leading back downhill to my tent, and then along the lake beyond the base of the glacier and into the canyon below, I thought about the paths that this and other glaciers have taken, and about the impressive forces of ice and the beauty of the outsized art it creates.

Though ice still covered Dana Lake in June, meltwater in dazzling turquoise had ponded on its surface. The glacier lies at the rear right, beneath the nearly vertical, snow-filled Dana Couloir.

CHAPTER TWO

CREATION OF A GLACIAL LANDSCAPE

Geographers, geologists, and glaciologists define a glacier as "a mass of moving ice created by the accumulation of snow." At the root of it, all that's needed is for more snow to fall than melts. As depth increases, the snow gets denser. Air, which might constitute 90 percent of fresh fallen snow, is squeezed out by "vapor transfer" that turns the snowflakes into pellets, by the weight of snow and rain accumulating above, and by repeated melting and refreezing. The light crystalline fluff transforms to granular snow the consistency of corn, and then to ice in a progression of hardness and density. The whiteness of a drift turns to vitreous gray and then to blue in places where you can see into its depth.

Though apparently solid, the protean ice is somewhat plastic—individual crystals can slide over one another, and the entire mass can also move, more or less, as a unit, especially when lubricated by water at its base. Snow and ice pile on top of previous layers and the pressure causes "deformation" sliding, and gravity tugs the entire mass down the slope of a mountain. The steeper the grade and the warmer the weather, the faster the slide. The surprising outcome of glacial physics is that the ice flows like a very slow river all the way from its uppermost attachment to the mountain to the end of its path, where it breaks into pieces and melts. Thus, a stone that falls onto the ice at the top will ride the glacier like a boat floating on a river and eventually be delivered to the spot down below where the ice ends, all the while scarcely rolling a single turn on its own.

Bedrock, boulders, and rough terrain underneath the ice lie out of sight, but their influence is felt everywhere because they create resistance to the flow above them. This changes the rate at which the ice moves and also the angle of the glacier's pitch downhill, as if it were sliding down a staircase of irregular-sized steps interspersed with both landings and steep drop-offs. Conflicting stresses of expansion on the outside radius of bends and compression on the inside contribute further to the varying rates of movement and to the changing angles of the ice, and together these forces cause the glacier to fracture into cracks called "crevasses." These open to widths of inches, or feet, or dozens of yards on large glaciers. Wide at the surface, crevasses taper shut at the bottom, which can be one step down or 150 feet deep. A fall would be like going from the top of a fifteen-story building to a totally black basement via the elevator shaft rather than the stairs.

Crevasses have formed in this glacial remnant on the east flank of Mount Gibbs.

The roughness of rocks and landforms underneath the glacier results not only in crevasses, but also in the collision of adjacent ice masses moving at different speeds. Like a lineup of boxcars on a derailing train, chunks of ice collide and rupture into tortured piles and shapes. On large and active glaciers, these upraised, blocklike features are called "seracs." All this upheaval slumps as it melts, fuses together again, or pulverizes and blends as it continues to move, evolving chaotically as a migrating mass of icy cubes, trapezoids, pyramids, columns, towers, and pinnacles. Together these comprise an icefall—one of the sculptural marvels of the glacial world.

Large glaciers have shaped the landscape beneath them. They scraped the earth down to bedrock and, like conveyor belts, they carried massive amounts of rock within and on the ice as they migrated downhill, ultimately depositing their load as a valley-wide rockpile called a "terminal" moraine. These typically dammed the streams running down from above. Popular glacial-moraine lakes on the east side of the Sierra include Donner, Fallen Leaf, Grant, June, and Convict. Thousands of other lakes across California's more remote terrain were formed by the scouring of glaciers or the moraines they left behind.

In the same way that the glaciers deposited rock at their lower ends, they left it at their sides as "lateral" moraines—levee-like ribbons of rock and dirt along the length of the glacier. In some places these became minor mountains in their own right. Some can be easily seen today, lining the sides of lower-elevation canyons in the eastern Sierra, including Green, Lee Vining, Bishop, and Big Pine.

Climber Josh Helling looks into a crevasse of unknown depth near the top of the Maclure Glacier.

Seracs have piled up among a sea of crevasses in the icefall of the Hotlum Glacier on Mount Shasta.

Rocks and boulders of many sizes fall onto the Hotlum Glacier and are then carried on the conveyor belt of the ice as it slowly flows down to the terminus of the glacier.

Meltwater flows across the surface of the lower Hotlum Glacier. Runoff sometimes drops into holes in the ice called "moulins" and then flows as a stream underneath the glacier.

Rocks and granite boulders have been deposited here as a typical high-elevation moraine of the Matthes Glacier, which is now reduced to the shaded slopes beneath the Glacier Divide in the South Fork San Joaquin basin.

The glaciers also pluck rocks from the headwalls and sides of mountains. Those rocks freeze into the ice, and the glaciers indomitably carry them along on their down-bound paths. Meanwhile, the bottoms of glaciers catch exposed edges of rocks, thus quarrying boulders and rolling them downhill along the super-abrasive interface of ice and earth. This generates rock "flour," which then washes into streams, making them look milky or silty gray and later—once most of the suspended grit has dropped out—a lucid blue-green that signifies distant glacial origins.

As rocks accumulate on top of the glaciers and snow diminishes in a changing climate, some ice glaciers become "rock glaciers." The deep layer of rocks tends to insulate the ice from further melting. These fascinating hidden features store water, affect microclimates, and, where ice melts on hot days and seeps out, create wetland habitats. They are common in the high glacial basins ("cirques") of the Sierra and will likely be the last glaciers surviving there.

North America has had three types of ice glaciers. First, "continental" glaciers swept southward from northern climes during the ice ages, last peaking about twenty thousand years ago, when the climate was 7 to 13 degrees Fahrenheit colder than it is now. These vast ice-scapes covered half of North America at the height of glaciation and buried the northeastern and north-midwestern states. Many times over a 2.5-million-year span, these mega-glaciers formed and melted in cycles lasting about one hundred thousand years each. They shaped much of the landscape we see today in New England and the upper Midwest. Through each of their cycles, they scraped soil and exposed bare rock, left narrow hills called "drumlins," truncated streams and created waterfalls, and blocked rivers. For a time, ice damming the outlet of the Saint Lawrence—second-largest river on the continent—shunted the flow of the Great Lakes down the Hudson River instead. In the Northwest, ice dams redirected the entire Columbia River across eastern Washington. Tens of thousands of lake beds were gouged out and still dominate the landscape in northern Maine and Minnesota, along with much of Canada. For hundreds of miles the continental

Lateral moraines of an ice-age glacier that once flowed down Green Creek Canyon, south of Bridgeport, curve through the center of this snow-covered scene in January. No glacier is present now, but the entire shaded canyon area was once filled with ice. The east-side Sierra glacier created the lateral moraines when it deposited rocks alongside its path.

glaciers deposited moraines along their southern edges; Long Island, New York, is one of these.

The continental glaciers held enough frozen water to lower the oceans three hundred feet below today's level. They pulverized and rearranged so much rock and grit that the resulting windblown "loess" piled up as deep fertile soil across the Midwest. But these ice masses did not reach California, owing to the weak effect of polar air currents there, the semi-arid climate to the east, and mountain-range barriers to the north.

California's counterpart to the continental ice masses was the Sierra ice cap, blanketing the highest elevations for an impressive length—a hundred miles—and oozing downhill to fill hundreds of canyons with the second type of glaciers, "valley" glaciers. Characteristically long and narrow, these formed an extravaganza of crevasses and seracs, and after melting, they left stair-steps of excavated landscape. The big valley glaciers cut through their terrain faster than the smaller tributary glaciers, and so after all the ice melted, the tributary streams were left high in "hanging valleys" to plunge into the receiving valley over waterfalls. These glacier-created falls are a highlight of any trip to Yosemite today.

Because their ice filled the valley floors, valley glaciers sculpted their canyons into U-shaped cross sections. In contrast, erosion by water alone forms V-shaped cross sections; water's erosive power is focused at the lowest point where it flows. The broad, open, relatively flat-floored canyons and valleys carved by glaciers create the scenery that we find so grand in Yosemite Valley, as well as Cedar Grove in Kings Canyon National Park. The upper North Fork of the Kern has a remarkable, rarely seen U-shaped trough where ice followed the canyon along a ruler-straight fault line.

California's valley glaciers flowed down the west slope of the mountains along the path of virtually every major river from the Yuba to the Kern, and also down many of the shorter east-side canyons. Ice covered the upper ten miles of the North Yuba basin down to today's Sierra City, the South Yuba to Emigrant Gap, and the South Fork American to Twin Bridges. The Tuolumne Glacier, including today's remnants—the Lyell and Maclure Glaciers in Yosemite National Park—was four thousand feet thick. It excavated the Grand Canyon of the Tuolumne River and continued as California's longest glacier for sixty miles, down to the 2,000-foot elevation. The similarly grandiose Merced Glacier flowed through Yosemite Valley. The Truckee Glacier halted abruptly at the Lake Tahoe outlet, where glacial dams six hundred feet tall were formed several times. The ice advanced down some east-side canyons the whole way to where Highway 395 now crosses the respective drainages. Other glaciers formed on a few isolated landmarks in the Klamath Ranges, at Lassen Peak, and on Mount Shasta. All of the Sierra Nevada's valley glaciers melted completely away in an interglacial warming period that began ten thousand years ago.

North America's third type of glaciers are "mountain," or "cirque" glaciers. Limited to high terrain, these are

A glacier and its cirque lie between peaks at the southern end of the Minarets. The upper end of the glacier has "plucked" rock from the mountain wall to create a carved-out basin and high pass, now filled with fresh snow on top of ice.

moraines. Where the huge volume of rock was "plucked" by the glaciers, giant depressions remained. Cirques are diagnostic of past glacier activity, and virtually every stream in high mountain terrain of the Sierra Nevada begins in one. You see them most pronounced on the north and eastern slopes, because these are the cooler aspects, where glaciers grew best. After the ice melted, the lowest parts of cirques often filled with water to create the stellar lakes and postcard-perfect views that define mountain scenery for many people.

In the United States, significant mountain glaciers

These yellow *Podistera* are the highest wildflowers to gain a foothold on moraines at the base of the Middle Palisade Glacier.

the glaciers we see today. They've typically formed just below ridgelines where snow blows over to the leeward, or downwind, side of the mountain crest, and where avalanches pile deep.

The ice-carved cirques are the most durable and ubiquitous relic of the mountain glaciers. These excavations resemble bowls with one side open and spilling to the slope below. They were formed where glaciers met the mountain walls, and water froze in the cracks and then expanded and popped rocks free to travel downslope to

Early morning light shines on the Mount Dade bergschrund—the gap where the ice separates from the mountain in the summer.

Bergschrunds can open to unknown depths where falling rock and ice are a constant hazard.

The Minarets rise with their flanking glaciers above a lake appropriately named Iceberg—one of many high-country lakes excavated by ice-age glaciers and dammed by moraines.

Oval-leaved buckwheat is one of the early colonizing plants on the rocky soil that the ice leaves in its wake. This flower grows in the glaciated saddle between Mounts Dana and Gibbs.

The Conness Glacier scraped the upper basin of Lee Vining Creek down to bedrock during the Pleistocene, leaving a chain of stellar mountain lakes. What remains of the glacier is tucked against the summit wall.

can still be found in the North Cascades, the volcanic peaks of Washington and Oregon, and the Olympic Mountains. The Wind River Range of Wyoming has the largest glaciers left in the Rockies. Glaciers can also be found in the Tetons of Wyoming, the Colorado Rockies, Glacier National Park and the Beartooth Plateau in Montana, plus the Sierra Nevada and Mount Shasta. In a class by itself, Alaska has many mountain and valley glaciers, and vast ice fields that resemble the continental glaciers of the past.

The glaciers of California may have completely melted after the Pleistocene ice ages; however, hundreds of cirque glaciers were reborn in what glaciologists call the "Little Ice Age," from 1350 to 1850. These receded in the drought years of 1910 to 1934, then advanced temporarily, then receded again and have been shrinking at an ominously rapid rate in recent years, as are most other glaciers throughout North America and the world.

The remnants of California's glaciers are scattered across three mountain ranges, and they vary from a few surviving rivers of ice to token relics of an era that will soon end.

Glaciers on Mount Shasta are the largest in California and the most often seen—visible from Interstate 5. In this view from the east, the Hotlum Glacier lies on the right, the Wintun Glacier is on the left, and a large snowfield fills the center right.

CHAPTER THREE

SEEING CALIFORNIA'S GEOGRAPHY OF ICE

The ethereal sight of Mount Shasta was my first view of anything in California. From my base at Oregon's Crater Lake, where I worked as a seasonal laborer for the National Park Service when I was nineteen, I had hitchhiked south for the weekend. As I topped a rise on Interstate 5, Shasta arose in elegant upturned lines on the horizon with a mass, scale, and brilliance unlike anything I had ever seen.

Though I didn't know it at the time, the luminous beauty of that mountain came from the snow of its north-side glaciers—masses of ice gleaming at the height of summer. The snow seemed not just unlikely in 90 degree heat, but impossible. Yet it was real, bold, and striking. I've since tried to imagine what the great mountain would look like if it were all dark rock with no snow and concluded that the allure of this California landmark would surely be diminished.

Years later, along a different roadside, far more remote, at the eastern fringe of the state, I found a different panorama no less astonishing than that of Shasta. I rolled to a stop high on the western flank of the White Mountains, above the town of Big Pine, where the Bristlecone Road curves north. The crest of the Sierra Nevada, with its multitude of high peaks and granite pinnacles reaching to 14,000 feet, lay spread out before me. The mountains were serrated, indomitable, still rising. The summits were striking, but the whiteness of slopes beneath them caught my attention every bit as much. Great swaths of snow seemed vertical in the foreshortened view from that distance. These were the Palisades Glaciers, and the refreshing coolness of their snowfields was palpable on the hot afternoon; they shone above the gray rocky rubble rising from green forests, golden canyons, and sunburned plains.

Only a few other roadsides offer thin glimpses of the glacial landscape of California. Matterhorn Peak's enticing crags and glacial slopes rise above irrigated pastures along Highway 395 north of Bridgeport. The Minaret Summit view of Ritter Range glaciers is seen from a popular overlook along the Devils Postpile Road west of Mammoth. Distant ice clinging to Shasta's southern face can be spotted from Highway 89 east of McCloud, and a few flashes of white show

The white mass of the Palisades group of glaciers, seen here from the Bristlecone Road west of Big Pine, expands with snowfields in late spring. The Middle Palisade stands at the far left with the Norman Clyde Glacier to its right. The Palisade Crest is the jagged ridgeline in the center, with its glacier below. Mount Sill is the dark peak in the distant right, followed by the North Palisade (touched by clouds) and Thunderbolt Peak. The large snow masses to the left and right of Thunderbolt are the North Palisade Glacier. Temple Crag is the large dark feature below the horizon on the right.

Seen from the Minaret Summit Overlook west of Mammoth, snow freshens the Ritter Range in late spring. Glaciers lie under the snow in all the cirque basins beneath the abrupt gradient of the peaks.

on the Palisades peaks as you speed along 395 between Bishop and Big Pine. But that's about all you'll see from the car. To write this book, I wanted to see the glaciers better, and to reach as many of them as I could.

Altogether, some sixty to five hundred remain in California, depending on what you call a glacier. Based on his own explorations, John Muir counted sixty-five—an amazingly accurate figure considering that his enumeration predated human flight and the comprehensive view it affords. In the 1970s, geologist Mary Hill estimated that seventy survived in the Sierra. Casting a bigger net, in 1980 William Raub and others of the US Geological Survey finished the most complete inventory of Sierra glaciers to that date. Their work was based on 1972 aerial photos (but wasn't published until 2006). With little concern for minimum-size criteria, they reported 497 glaciers and another 788 ice patches. These ranged from 390 acres down to 2.5—the smaller ones likely long gone by now. Refining and updating this inventory in 2008, Hassan Basagic, a graduate student at Portland State University, started with a list of the 1,719 permanent snowfields shown on USGS maps and screened each to see if it had a minimum of 2.5 acres of perennial snow, plus the requisite mass to cause it to move downhill. This was estimated by comparing the area of each permanent snowfield to the size of glaciers worldwide known to be moving. He concluded that the Sierra Nevada likely has 122 glaciers with the "critical shear stress" that would cause them to

move downhill. With Shasta's ten glaciers identified by geologist Philip Rhodes in 1987, and two in the Trinity Alps, the total comes to 134 for California. Using somewhat different criteria, geologist Bill Guyton had arrived at a similar total, 108 (with 401 "glacierets") in his 1998 book, *Glaciers of California*.

All that said, there's no official inventory, and the presence of ice under snow and its movement over time can be difficult to spot in the field, or by any other means, for that matter. My rule of thumb is to look for a bergschrund. Though some amount of opening at the top of a glacier might owe simply to melting at the rocky headwall next to the ice, a bergschrund can usually be considered diagnostic of glacial movement: the ice has slid downhill, leaving a gap roughly the same size as the distance the glacier has traveled. But for years after the ice stops moving, the same ice is still there, and until it's melted and gone, it remains at least a remnant glacier to me.

By August, September, or October, most of the recent snow is melted off the glaciers, exposing ice underneath, and that's when you can see the size that they really are. In my photo expeditions, I didn't limit myself to these minimums but sought out the glaciers in late spring, when the deep snow that had dressed them in their formative season of winter was still visible and they most resembled what existed back when the glaciers were vast. I also photographed in midsummer, when snow still covered much of the ice with gleaming white. I especially photographed in early autumn, when most of the previous year's snow was gone and only the glacial ice remained, with its surface decorated by rock and dirt.

California has three glacial regions. First, in the Trinity Alps of northwestern California, one or two small glaciers appear to endure as the last in the entire coastal mountain ranges south of the Olympic Mountains of Washington. Second, Mount Shasta has ten glaciers, including the state's largest and most active. Third, and by far the largest region, the Sierra Nevada holds most of California's glaciers and remnant ice fields. All are on high north- or east-facing slopes, and most are in the central range between the north boundary of Yosemite National Park and the Palisades peaks southwest of Big Pine. North of that belt, elevations are too low; south of it, snowfall in most places is too light owing to the increasing effect of the North Pacific high-pressure system, which pushes snow-laden northerly storms away. Only small glaciers and icy remnants persist south of the Palisades, and most of those won't last long; the twelve-acre Lilliput Glacier hugs the north side of Mount Stewart in Sequoia National Park, and the seventeen-acre Picket Glacier lingers at the headwaters of the North Fork Kern.

The Palisades peaks have the greatest concentration of Sierra Nevada glaciers, and to see the southernmost of these in July, I took the Glacier Lodge Road west of Big Pine and hiked up the South Fork of Big Pine Creek. The trail ended at an ancient cirque where Brainerd Lake

mirrors the trees and rocks around it. On fainter paths I scrambled up to the stellar Finger Lake. Then beyond any trace of trails, I passed waterfalls and austere lakes and ended with a rock-by-rock approach over a mile-long moonscape of terminal moraines left by the Middle Palisade Glacier back when it was strong. At dusk I emerged on a rise of bare broken rock at the edge of the ice.

An array of silhouetted peaks rose above me. At my left, The Thumb towered as a horn that had been glaciated to nearly vertical slopes on all sides, followed by Disappointment Peak. Then the Middle Palisade soared to 14,012 feet and spawned a cirque glacier that's split in two. Marking the far right shoulder of this basin, Norman Clyde Peak towered even more impressively than the Middle Pal. I camped after moving rocks to excavate a sleeping spot, though I couldn't move too many because meltwater from adjacent snowbanks lay ponded underneath, and the cushion of sharp granite was all that kept me from being soaked in my sleeping bag from the bottom up.

On that midsummer evening at the base of the Middle Palisade I sat and watched the snow melt, the streams flow, the clouds pass, the night descend. The peaks capped the scene perfectly, and the glaciers lay tightly embraced and shaded by the ridges and summits around them. The wind eased, the light faded from gold to navy blue, and stars fell into the sky. In the hush of quiet that grew around me, rocks rattled from the ice's edge and reminded me of

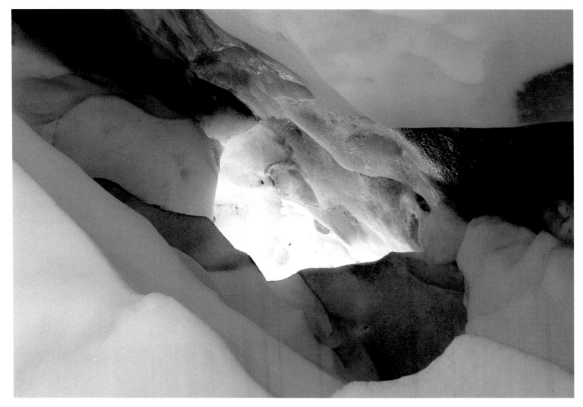

This summertime snow tunnel forms inside the bergschrund of the south lobe of the Middle Palisade Glacier.

the glaciers' power to reform whole mountains. The night grew colder, and the snow and runoff froze once again into crystals that would delay the passing of the glaciers for one more day.

The pictures that follow show some of what I saw at the Middle Palisade and on my other glacier expeditions, from northern California's Trinity Alps to the southern Sierra Nevada.

52

PORTFOLIO OF CALIFORNIA GLACIERS

Northwestern California's Thompson Peak rises above its glacier in July after a springtime of heavy snowfall. This 9,002-foot summit caps the high country of the Trinity Alps.

Slabs of broken cornice near the top of Caesar Cap Peak, immediately east of Thompson Peak, cover a glacial remnant at the headwaters of the South Fork Salmon in the Klamath River basin.

A glacier lies beneath the vertical rise of Three Teeth, a part of Sawtooth Ridge, southwest of Bridgeport. This is one of the Sierra Nevada's northernmost glaciers, seen here with heavy snow cover in July.

Snowfields and glacial remnants lie under Dragtooth Peak, just northwest of Matterhorn Peak.

Matterhorn Peak rises from its glacier south of Twin Lakes. It can also be seen in a distant view from Highway 395 north of Bridgeport. Whitebark pines grow closer to California's glaciers than any other trees.

At the northeastern edge of Yosemite National Park, Mount Conness rises above its glaciers and deep snowfields in springtime, seen here from the shoulder of North Peak.

In September the remaining ice of the Conness Glacier clings to the base of the 12,590-foot mountain.

The Dana Glacier lies under deep snowfields in early June. Though it once would have filled much of the background of this photograph, the ice is now limited to the area near the base of the couloir. This northeast face of the mountain is heavily excavated by the glacier, but Mount Dana is better known for its nicely rounded, unglaciated west-side profile, a backdrop to Tuolumne Meadows.

Facing page: The Mount Gibbs Glacier (top of photo) and its residual ice fields cling to the eastern escarpment of the Sierra in autumn.

Yosemite's Mount Maclure rises over its glacier in September, with snowfields drifted on its lower elevations.

In July, Banner Peak (center) and Mount Ritter (left) tower over Thousand Island Lake in a year of heavy snowfall. Several glaciers fill the cirque basins of the two central Sierra peaks.

The Clyde Minaret (right) rises from a frozen pond above Cecile Lake. Glacier remnants cling to the upper slopes in the background.

Ten discrete Minaret glaciers lie against the pinnacles of the Ritter Range, west of Mammoth Lakes, seen here at its northern end in July.

With a snow-covered glacier at the base of its vertical rise, Clyde Minaret catches the golden light of morning.

Mounts Abbot (left) and Mills (right) shade small glaciers at the headwaters of Rock Creek.

The Mount Dade Glacier and its bergschrund begin to warm with the day's first sunlight.

The lofty summit of Mount Humphreys, 13,986 feet, casts a shadow that supports a glacier and its adjacent snowfield to the right of the summit, seen here in July from the semi-arid Buttermilk Country west of Bishop.

Dwindling Goethe Glacier is shaded against the Glacier Divide in the Piute Creek basin at the headwaters of the South Fork San Joaquin River.

The thin line of the once-extensive Matthes Glacier is reflected in the still waters of Lobe Lakes, beneath the Glacier Divide, in September.

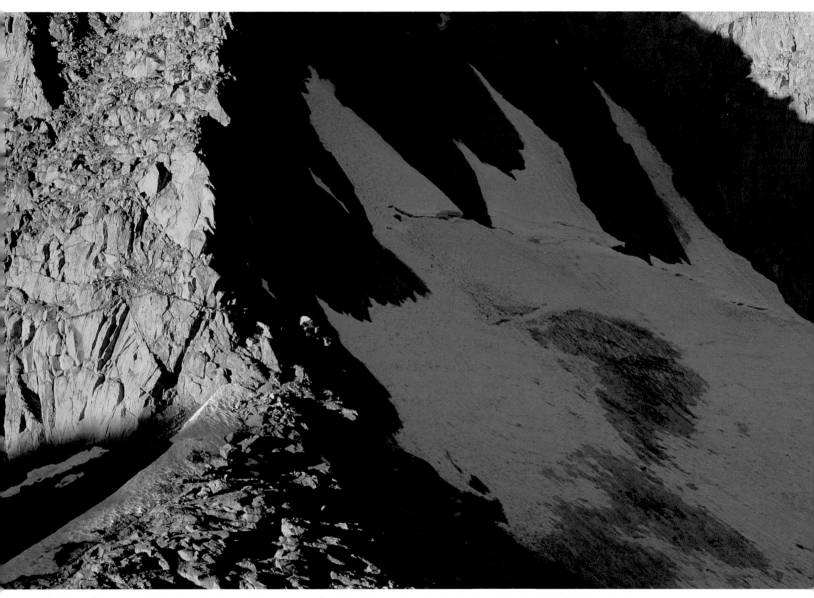

The Mount Thompson Glacier benefits from shade cast by a shoulder of the peak at the headwaters of the South Fork Bishop Creek.

The Mount Gilbert Glacier, just east of Mount Thompson, has been reduced to a thin veneer of ice, mostly covered by rock in September.

North Palisade (far left), Thunderbolt Peak (center left), and Mount Winchell (distant far right), west of Big Pine, are all whitened with fresh snowfall in late May. The North Palisade Glacier—largest in the Sierra Nevada—lies up against the high peaks on the far left and far right.

The north lobe of the Middle Palisade Glacier lies below its namesake summit and Norman Clyde Peak, which rises on the far right.

In a time exposure taken at one a.m., moonlight illumines the Minaret Glacier and the Clyde Minaret.

Nettie Pardue pulls in the slack of her safety rope before going any farther on the icefall of the Hotlum Glacier on the east side of Mount Shasta.

CHAPTER FOUR

MOUNT SHASTA

On the towering stratovolcano of the north, California's largest surviving glaciers still move, rupture, creak, crack, groan, push rock, and swallow the unwary. For now, you can still go out and find glacial features that look like they might have looked in the ice ages.

I was eager for this refreshing scene.

With great expectations I went to the 14,162-foot Mount Shasta at the end of August, when I thought that most of the previous season's snow would be melted away, and before new snow was likely to fall. But you never know. When I arrived at the end of the road above Shasta City, the mountain's signature cloud cap hovered over the top like a lens. By dusk thicker clouds were lowering all around me, and by dark they had settled like fog. Sometime in the night it began to drizzle. Then it rained, poured, sleeted, and snowed for two days and nights while I waited in my van with the propane heater kicking on and off. In summer weather like this, I had plenty of time to imagine glaciers growing as they did in the past but rarely do today.

During my third night up there, the drops on the roof grew quiet. I checked the sky at four a.m. and saw both stars and clouds. Fearing that the window in this autumnlike siege of weather might be narrow, I dressed in a rush and dashed out before dawn, headed for the mountain's nearest glacier. It was nestled into a ravine at 11,000 feet, 3,000 feet above me. I suspected that the usual hiker's route would be badly iced where it crosses steep terrain at the base of a looming crag called Shastarama, so I walked on fields of basalt to the north and climbed for two hours directly up a steep 2,000-foot slope of volcanic rock that faced southwest and so didn't hold much new snow. Not long after the sun crested the eastern horizon, I emerged at a ridge where the Mud Creek Glacier spread gloriously below. The sun shone through a moist atmosphere, but the wind began to howl as I walked across the wintry, snow-covered glacier. I rounded its crescent curve to the east, staying on rocks at the edge, just in case the new snow had hidden a crevasse. Soon the clouds socked in again. Flurries began to fly, then increased, then thickened. Though in the grip of the late-summer blizzard, I felt grateful to see snow piling deeper on the mountain's southernmost glacier.

Mud Creek Glacier curves to the north and east while an early morning squall begins to blow. The base of Shastarama Point stands on the right.

My unseasonable introduction served as a warning that you can pay dearly for mistakes on this mountain, and I sought advice back down in Mount Shasta City from Jenn Carr at Shasta Mountain Guides and from local climber Leif Voeltz at his store, The Fifth Season. They helped me hatch a plan to see the best of the glacial landscapes left in California. Climbing alone on Shasta's larger glaciers is not advisable, so I called two friends who were standing by in San Francisco, and we agreed to meet three days later at the portal to the Hotlum Glacier.

Groceries bought and water jug filled, I drove off toward Shasta's northeastern flank. A two-hour maze of dirt roads took me to the obscure Brewer Creek trailhead at dusk, just in time to stroll up through the red fir forest for a red-sky view of the mountain, glowing in its mantle of snow and ice that still looked fresh and white even at the end of August.

Unlike any other mountain in the state, Shasta rises both north enough and high enough to get plentiful snow and minimal rain—an increasingly rare combination at California latitudes. In contrast, Lassen Peak, which is Shasta's nearby companion and the southernmost Cascade Range volcano, looks impressive but lies nearly 4,000 feet lower and so lacks surviving glaciers. Furthermore, snowfall on Shasta has been increasing with the intensified storms spawned by the warming ocean temperatures associated with climate change. Glaciologists have found that, unlike 90 percent of the glaciers worldwide, some of Shasta's glaciers are not shrinking, but rather thickening— at least for now. The small south-side glaciers have roughly doubled in length since 1950.

However, experienced climbers know of places where the ice has markedly receded, and with global temperatures expected to rise 4.7 to 10.5 degrees Fahrenheit by 2100, according to the California Climate Change Center, summertime warming will eventually cancel out the added snowfall or turn it into rain. Even the Shasta glaciers are likely to lapse into rapid retreat during this century, according to glaciologist and Shasta researcher Slawek Tulaczyk at the University of California, Santa Cruz. This is no casual loss: snowfall and glacial melt supply the Sacramento River, which provides water to most of the state's cities and farms. And the glacier-encased north side of the mountain sends chilled runoff through underground cavities to the Shasta and Klamath rivers, where cool flows nourish important runs of threatened salmon and steelhead.

Contributing much of that flow, the northernmost and longest of the ten Shasta glaciers is the 2.4-mile-long Whitney Glacier. This is the only remaining valley glacier in California. It lies not only in the north-side shade of the mountain, but also in the shadow of Shastina. At 12,330 feet, and attached like a big limpet to the west side of the larger peak, Shastina is the third-tallest volcano in the entire Cascade Range, taller than all but Mounts Rainier and Shasta. Crevasses decorate Whitney's upper slope, but then the glacier flattens and carries a moveable landscape of volcanic rock. Finally the ice terminates at a blackened sea of basalt where raging outburst floods of snowmelt, mud, and rocky debris occurred in 1985 and 1998, eliminating a trail.

Other substantial Shasta glaciers are the Bolam facing north, the Hotlum facing northeast, and the Wintun facing east. The Chicago Glacier is a lower lobe of the Hotlum. On the mountain's south side, the Konwakiton clings to high slopes and is nourished by windblown snow and massive cornices that collapse onto the ice. Watkins survives as a bean-shaped patch of ice facing east, and the Mud Creek Glacier endures owing to Shastarama Point and the shade it throws onto the glacier's basin. Geologist Philip Rhodes also identified the Upper Wintun and diminutive Olberman Glacier northwest of Mud Creek.

Shasta's largest glacier in surface area is the two-square-mile Hotlum, and aiming to acclimatize for a few days at its base, I set out to explore the lower levels the next morning. In glowing sunshine I started up the trail and then aimed directly toward the Shasta summit. Wind-sculpted groves of whitebark pines thinned, and then scree slopes slanted steeper and steeper. I emerged at a small lake that was silty with rock dust. Above it I negotiated a crumbling moraine and then finally wandered the crusty, rock-riddled surface of the Hotlum Glacier. It was broad and hard, impressive with corrugations and the algid wreckage of centuries. Boulders as big as desks patiently rode the ice intact, along with fragments of every smaller size. Finding it impossible to camp on ice veined by flowing rivulets of water, or on the adjacent naked rockpiles, I dropped back down to a hospitable flat below the glacier at the head of Gravel Creek. After returning to the ice at daybreak for photos, I headed back to my van to await the arrival of my friends.

Jeff Pflueger and Nettie Pardue have been climbing, exploring glaciers, teaching mountaineering skills at Outward Bound, taking pictures for a living, and making a life of adventure for years—perfect partners for my Hotlum trip. After negotiating the long sequence of dirt roads, they arrived somehow smiling at one a.m. The next morning they shouldered bulging packs with climbing gear that dwarfed my load of camping and camera equipment, but they didn't mind the slightest. Later earning my envy, they seemed

Carrying rocks along with melted and recomposed remains of seracs and crevasses from far above, the Hotlum Glacier flows down to its end.

perfectly impervious to altitude headaches as we trekked for six hours up the Hotlum-Wintun Ridge, which was largely melted out between two of Shasta's major glaciers.

Halfway between the trailhead and summit, at the 11,600-foot elevation, we camped at a lull in the ridgeline's otherwise craggy rise where we could cram two tents among the boulders. Quite an eyeful out the door, a hundred-foot cliff dropped off directly to the glacier. Riddled with gaping crevasses, the ice had heaved and collided with itself on the downward journey, leaving large areas as inaccessible as sinkholes. Chasms, overhangs, and tipping seracs filled our view while Jeff melted snow for us to drink after dinner. Sinking into a deep chill as the sky darkened, we zipped up our tents and sleeping bags for a long, windy, frigid night.

In the morning we collected our gear, found a ramp to the ice, strapped on our crampons, and tied ourselves into a fifty-meter rope with butterfly knots. The purpose of the rope was to enable two of us to catch the fall of the third person should something go terribly wrong, such as the collapse of an ice bridge over a crevasse. We clutched our ice axes as if our lives depended on them, and started across the glacier.

There at the Hotlum icefall, the glacier's movement combined with obstructing topography and accumulated snow hundreds of feet deep had produced gorgeous chaos. Imagine a small neighborhood where all the houses have been turned to ice, then shaken until they broke into fragments of all sizes, and then dumped out there in the storms. What might under other conditions have been a smooth,

Morning light strikes the middle icefall of the Hotlum Glacier.

skiable surface was evenly split between features that rose up above us—shear walls, leaning towers, crumbling facades, and blocks in every precarious state of balance—and those that lay beneath us—cracks, crevasses, caverns, open vaults, and spooky pits of empty air with water at the bottom. Shallow depressions in the ice were predictably white. Deeper cuts were tinged baby blue. Still deeper cuts had the beguiling color of the ocean at Big Sur, and the very deepest were navy blue—the color of an almost-dark sky at twilight, which was somehow disquieting there in the middle of the day. On our visit these colors were muted because the early snows, including the six-inch dump from only days before, had rewhitened the autumn scene. A few cavities nonetheless turned perfectly black inside narrow

Great seracs rise over the Hotlum Glacier.

openings. I made sure that Jeff and Nettie tensed the rope before I walked near these.

I was happy to have this virtual safety net, but I suffered no illusions: to arrest such a fall is no easy feat. If more than one of us fell, we'd all be toast, so to speak, in an icy pit, though Jeff or Nettie could presumably climb out with the use of ice axes and ice screws for safety anchors. In the best of several sobering scenarios, the two people left on top of the ice must know immediately that the third partner has disappeared, and at that instant lie face-down away from the chasm, slam their ice axes into the snow, and hold on for dear life while the rope snaps tight and the weight of the careening person is suddenly transferred to their harnesses. The victim might land on the bottom or on a ledge, relieving tension on the rope, or they might not, and be left dangling, their full weight on the rope-team above. Either outcome requires a skilled response. The rescuers would anchor the rope with ice screws or snow pickets or simply an axe, if that was all they had, then relieve tension from their bodies so they could move, and finally begin the rescue—possibly by rigging a pulley system to hoist the victim up. Jeff and Nettie know how to do this: they have practiced, caught teammates, and have even been caught by others in surprise falls into Alaskan crevasses. I had done a bit of book-learning as I prepped for my summer adventures but—being out there by myself most of the time—had never rehearsed such a high-consequence drill.

Ice formations point to the sky and threaten to fall in the warming sun.

"Don't worry," Jeff said. "We'll take care of you. That's why you're in the middle of the rope and we're at the ends. If one of us falls in, just self-arrest with your axe, and we'll tell you what to do. If you fall in, we'll still tell you what to do."

On glaciers with hazards, which we undoubtedly had, two entire roped crews of at least three people each are advised. So we didn't take any chances. Much as Jeff wanted to traverse an awesome snow bridge arching over a

Ice towers are backlit with afternoon sun. The rock in the foreground is underlain by glacier.

Climber Jeff Pflueger waits on top while I photograph at the base of towering seracs.

end. Because we still have glaciers in California today, we can still see all of this and imagine the great forces and the great creation story of the past.

From our vantage point that day in the blue light beneath Shasta's walls of ice, it seemed as though the climate and the landscape and the life that we have long known would continue to function under the greatest working principles of the natural world—timeless rules that govern the oceans and the atmosphere, the evaporation and freezing of water, the avalanches, the flow of rivers, the growth of forests, the migrations of birds. But that world is changing.

hundred-foot gulf for an ice climb of a lifetime, Nettie said no, and I voted with her.

Surrounded by the rise of ice towers and the openings of crevasses, we stopped to throw snowballs in and listen for depth beyond what we could see. Long seconds elapsed before we heard a splash.

We had crossed the threshold to a world apart, to a world of beauty unlike any other I had ever seen. We had entered not only a place but also a process that has shaped some of the most exquisite landscapes in California, from the slopes there at Shasta to the upper canyons of the Kern; from the Great Basin, where eastern Sierra ice fields once terminated at the edge of the desert, to the western foothills, where the Tuolumne Glacier finally crunched to its

On the Hotlum Glacier Jeff has found a serac stable enough for ice climbing.

Greg Stock crosses the top of the Lyell Glacier to take measurements of the diminishing quantity and flow of ice.

CHAPTER FIVE

THE MESSAGE OF THE GLACIERS

Greg Stock set his pack down in the snow, opened the flap, dug inside, and pulled out a global positioning device to pinpoint his latitude and longitude down to the inch. The Yosemite National Park geologist had already found a benchmark that was established on the bedrock margin of the Lyell Glacier in 1949 for the purpose of measuring the height of the ice surface. Now he mounted the GPS unit onto a backpack and started walking across the glacier.

With the help of Steve Bumgardner, who was making a film about glaciers for the Yosemite Conservancy, and Josh Helling, a notorious climber and the expedition's safety expert, Greg had already pinpointed a similar benchmark on the far side of the ice. When these markers were installed sixty years ago they were flush with the ice surface; now they were perched high above. Greg's measurements would tell how much the glacier had thinned over the past sixty years.

The Lyell Glacier lies tucked into one of the deeper recesses of Yosemite National Park. It's not visible from roads but can be seen in the alluring view from the top of Lembert Dome in Tuolumne Meadows, and from the Pacific Crest Trail as it climbs Donahue Pass, south of the meadows. The ice slants down beneath the ridge cap of Mount Lyell, at 13,114 feet the park's tallest peak, and ranks as the second-largest glacier in the Sierra Nevada.

In collaboration with scientists from the University of Colorado at Boulder, Greg's research gauges not only the dimensions of the ice, but also its rate of movement and the runoff from glacial streams. Repeat-photo documentation, systematically done Sierra-wide by geographer Hassan Basagic in 2004, and for the past twenty years at this site by Yosemite Conservancy naturalist Pete Devine, shows that the glacier's surface area is dramatically shrinking. As soon as I arrived with Greg and his crew in September, I took a look for myself by pulling a copy of a 1903 photo from my pocket and comparing it with what I saw. Much had melted. Recent evidence, including Pete's photos, indicates a greatly accelerated decline in just the past decade.

And it's worse than that. Imagine a hypothetical case in which, each year, several feet melt from each side of the

The Mount Gilbert Glacier has shrunk to a thin, rock-stained veneer against a steep headwall. Small glacial remnants are likely disappearing from the Sierra Nevada nearly every year.

glacier, causing its area to shrink. That would show on the photos. But the surface of the glacier can also melt, causing the ice to thin. While still apparently large, the glacier could theoretically dwindle to a thickness of only one foot. The next year it would suddenly all be gone. Scientists believe that some version of this scenario is, in fact, unfolding; the demise of Sierra glaciers may be occurring even faster than the photos indicate.

After reaching the far side of the glacier, Greg recorded numbers from the GPS unit, calculated distances and elevations, and announced, "One hundred and twenty feet." That's how much the glacier had thinned in sixty years. The height of a twelve-story building!

It immediately raised the question, How much ice was left? Greg didn't know but feared it could be as little as tens of feet across much of the surface. With two feet of thinning per year on average, it doesn't take a lot of math to imagine the future here.

Throughout the Sierra, glaciers are about half as large in area as they were a hundred years ago, as reported by Basagic in the Glaciers of the American West project directed by Andrew Fountain at Portland State University. The average temperature in the High Sierra has increased 5 degrees Fahrenheit in that interval. By 2090 the temperatures in Yosemite are expected to be 7.5 degrees higher than they were in 1990, according to the Rocky Mountain Climate Organization. So the ice will continue to melt. The only region in the US losing ice at a faster rate

is Glacier National Park in Montana. A century ago, 150 glaciers inspired the naming of that great park; today only 25 survive, and all are expected to expire by 2020. The Wind River Range of Wyoming has the largest concentration of glaciers in the Rocky Mountains. Most will likely be gone in fifty years.

The century-long loss of ice has in the past two decades increased to a "dramatic retreat," according to the Portland State University study. The glacier of John Muir's first discovery, on Merced Peak, is gone. The Mendel Glacier at the San Joaquin headwaters used to be one of the finest ice climbs in the mountains, according to longtime Sierra aficionado James Wilson of Wilson's Eastside Sports in Bishop. "Now it's just a skinny slot of dirt-covered slush," he says. Even the relatively large Lyell appears to lack the mass and the accruing annual weight to cause substantial downslope movement.

The hour was late and the news was bad. Greg, Steve, and Josh headed down on a two-hour hike to the upper Lyell Fork, where their camp had been supplied by mules carrying a weighty cargo of technical and video gear. I dropped to the bottom of the glacier and pitched my micro-camp along a lakeshore at the base of the upper moraine, positioned to catch a sunrise photo.

The next morning, after my six a.m. shoot, I moved my camp a mile or so, to the base of the neighboring Maclure Glacier, and awaited my friends' return.

John Muir had measured the migration of this glacier

Above: Lyell Glacier's western lobe on September 14, 2010, reflects early morning light. Comparison to a 1903 photo shows less snow on the left and right sides of the glacier.

Left: This photo of Lyell Glacier was taken by G. K. Gilbert of the US Geological Survey on August 7, 1903, during his investigation of Sierra glaciers.

Above: The Darwin Glacier on September 25, 2010, appears on the left side of this photo and has separated into two lobes with much less snow and ice than appear in a 1908 historic photo. The Mendel Glacier lies to the right.

Right: This historic photo of the Darwin Glacier was taken by G. K. Gilbert on August 14, 1908.

Above: The Dana Glacier lies at the base of the steep Dana Couloir, at center right, on September 12, 2010. A historic 1903 photo shows an ice field far more vast and connected to the two distant couloirs in the background.

Left: This photo of the Dana Glacier was taken in 1903 by G. K. Gilbert. With crevasses at its side, the glacier spread across the wide cirque basin above Dana Lake.

with sticks placed into the ice in 1872 and found movement of one inch per day. Repeating the experiment, Greg now handed a reflector to Josh to place on PVC stakes in the ice. Then he measured the distance and angle between his position and the stakes, compared the numbers with those taken the previous year, and with a bit of elation in his voice, shouted up to where the rest of us waited, "Twenty feet!" Unlike many other remaining Sierra glaciers, Maclure was still flowing downhill at a substantial rate.

Like John Muir a hundred and forty years before, Greg had garnered hard evidence of "living glaciers" in Yosemite, but the greater bulk of his evidence confirmed that the glaciers were shrinking markedly year by year. Heavy snows of the previous winter did nothing to change the long-term trend; anyone looking to one or a few big winters as evidence that global warming is bunk may as well assume that darkness will never come just because the sun is high at noon. Ironically, 2010 was also the hottest year on record worldwide.

Greg seemed to take the message of the Yosemite glaciers personally, the way many of us might regard the prognosis for the value of our own homes or neighborhoods in declining economic times. The reason for this is simple: Yosemite *is* his home.

He grew up in the Sierra foothill town of Murphys, not far north and west of Yosemite, and visited the park as a child, always with the impression that it was protected and would stay about the same. But near his home he also witnessed the damming of the once wild Stanislaus River, and he knew from an early age that nature and wildness are fragile, their fate uncertain and subject to the whims of people. In college, he realized that development pressures at the edges of the park affected its wildlife, and that air pollution from far away killed or crippled many trees here. These experiences eventually led Greg to a degree and career in geology, including Ph.D. research on Sierra caves along the now-flooded Stanislaus. Then he moved to Alaska for fieldwork on the massive glaciers that remain—for now—in the Chugach and Wrangell Mountains. In 2006 Greg was hired as the first official geologist in Yosemite National Park, and with a combined sense of gratitude and stewardship, he returned home.

"A lot of my job involves issues of rockfall and public safety," he said of his responsibilities. Most of his work is in the heavily visited Yosemite Valley, where rocks, as a matter of course, fall from cliffs and land in places where people, as a matter of habit, walk, drive, camp, and lodge. His chief challenge is to predict what rocks and gravity—abetted by the inevitable storm or earthquake—will do in such places as the Highway 140 corridor and Curry Village. Greg recognizes the importance of that task along with its economic, legal, and administrative entanglements, but, he told me, "As a change of pace from the Valley floor, I really love the glacier work."

"The upper Tuolumne River and its ecosystems definitely benefit from the late-season flows provided by these glaciers," Greg added. The streams trickling out from the

A glacier lies in the background within the saddle between Mount Ritter, on the left, and Banner Peak, on the right. The headwaters of Shadow Creek come from this and other glaciers and provide important late-season flows to the San Joaquin River.

ice accumulated everywhere we walked at the base of the glaciers. Nearby drainages lacking glaciers were dry, their snowmelt gone by July. Farther south in the Sierra, the September flows of Los Angeles-bound Big Pine Creek, which has the range's largest cluster of glaciers, measure four times greater than those of Cottonwood Creek, which is alike in size and location but lacks glaciers. Fish, wild-life, and whole aquatic ecosystems depend on the flows, as well as on microclimates around the glaciers themselves.

Greg added, "The disappearance of these glaciers will likely cause changes that we haven't even thought about, and the suite of global warming problems might be the clearest illustration that the health of the park depends not just on what we do here, but on what happens all around us."

Worldwide, atmospheric carbon, which is the chief culprit in global warming, has increased by 36 percent during the past 250 years and will continue to grow, according to the United Nations Intergovernmental Panel on Climate Change in *Climate Change 2007*. For the rise to stop, or even slow down, dramatic adjustments in human behavior will be required, such as burning less fossil fuel and stopping deforestation. Even though scientists and informed people have known this for three decades, our society—geared to the greater engines of economic power and politics, ruled respectively by advertising and vested interests—has failed to address the human causes of climate change. Congress is still stuck on whether or not it exists; oblivious to an overwhelming consensus among climate scientists, a substantial number of congressmen—including one-half of the incoming Republican freshmen in 2011—deny that climate change is real. Elected officials claiming to be in denial just happen to reap large cash contributions from the oil and coal industries. Consequently, our lawmakers have barely considered what to *do* about the problem.

For the same reasons that the glaciers are fading, the Sierra snowpack is slated to decline, a lot, and this means really big trouble. The 4.7 to 10.5 degree Fahrenheit statewide increase in temperature predicted by the end of the century corresponds to an increase in elevation of several thousand feet in the snowline of a typical Sierra storm. Owing to the rising temperatures, scientists at Scripps Institution of Oceanography have projected a 30 to 70 percent or greater reduction in Sierra snowpack by the end of the century, even if corrective measures are taken. Other scientists reported in the *Proceedings of the National Academy of Sciences* in August 2004 that snowpack could be reduced by as much as 90 percent. This snowpack accounts for 65 percent of the water that people and farms consume in California; much of the future runoff will be arriving too early in the springtime to be usable. The California Department of Water Resources estimated in its 2005 *State Water Plan* that the change will reduce runoff from April to July by 52 percent and that farms could lose a quarter of their water supply. On top of all that, population at recent rates of growth would double in only forty to fifty years; even at the present reduced rate of growth,

California still grows by about four million people per decade. When the current problems are coupled with that kind of population growth, a challenge of truly daunting proportions lies ahead.

Climate change will have severe economic consequences. How will the engine of California run without its requisite supply of water? What will happen to coastal communities, real estate, and the entire shoreline when sea level rises 2 to 4.7 feet this century, as projected by the Climate Change Center? How will cities cope with increasing floods along rivers, intensified coastal storms, and heat waves that will make the scorchers now endured in Fresno or Redding feel like nice spring days?

From a purely nature-focused view, the changes written in the glaciers are tragic and, I might add, heartbreaking. As the climate warms, pikas—plump little short-eared relatives of rabbits—require cool habitat and so move farther upslope. But they lived at high elevations to begin with and eventually will reach the summits or near-lifeless rock piles and have nowhere else to go. Already they've disappeared from 30 percent of their previous habitat. The alpine chipmunk is similarly challenged. Snowshoe hares that I see above timberline, the elusive wolverine, and the common marmot all make their homes in the specialty niches of the alpine zone, and many of these creatures have no alternative.

Showy black-and-gray Clark's nutcrackers depend on the high-elevation seed crops of the whitebark pine, as do many other birds and animals; the tree is a "keystone"

The white-tailed ptarmigan lives at high elevations of the Sierra Nevada and depends on a cold climate that is now warming dramatically.

species. In the Rocky Mountains, grizzly bears feed on caches of these pine nuts, which the jays conveniently bury each autumn. But 77 percent of the whitebarks throughout the West appear to be succumbing to climate-related diseases. This is the first widely distributed tree to be nominated as an endangered species. Similarly, the oldest living organisms on earth—the bristlecone pines just across the Owens Valley from the Sierra—are threatened and declining owing to an exotic fungus and to bark beetles abetted by the warming climate. Some of these trees have lived for over four thousand years. Throughout much of Yosemite and Sequoia National Parks, the death rate of native trees has doubled in the past two decades. Worldwide, an estimated one-third of 1,009 mountain bird species will be "severely threatened" owing to climate change, according to Yale University researchers.

When I see the melting of the glaciers, I'm reminded of the projected demise of all these magnificent plants and animals, not to mention a hotter world of greater scarcity and increasingly harsh competition in our civilization down below.

"One thing we can do at the national park is understand the changes that are occurring and make those changes known," Greg said, reflecting on the enormity of challenges ahead. "As scientists and interpreters, I think we have that responsibility."

The challenges may appear to be overwhelming, but there's no shortage of doable tasks that must be accomplished, and the urgency is now recognized not only by environmental organizations but also business associations, churches, farm groups, and even the Department of Defense, which in 2004 concluded that our national security will be affected by the changing climate and resulting scarcity of water, resources, and places for people to live.

Sensible solutions to climate change include using less fuel and electricity through efficiency measures, recycling, protecting old-growth forests because they sequester carbon, planting trees on cutover lands, taxing industries and producers who unleash the most greenhouse gases, increasing our use of solar power, wind power, and other sources of renewable energy, redesigning American communities to encourage pedestrians and the use of mass transit, and stabilizing American population growth. With only 5 percent of the world's population, the United States now generates 25 percent of the greenhouse gas emissions that cause climate change, which is bad enough. But at recent rates of growth, our nationwide population will double in only about sixty years, in effect producing *twice* the burden on the rest of the world. The ambitious but challenging goal of reducing carbon emissions fades to a near impossibility when the current rate of population growth in the United States is considered, so coping with that growth curve should be at the top of anybody's "to-do" list regarding climate change.

The glaciers have come and gone before, but today this loss is different because now we are the cause of it, and because it is happening very fast. It's one thing for the climate to change within great cycles of the earth's orbit, but another for it to change because of avoidable pollution and deforestation. It's one thing to see the glaciers come and go in cycles of hundreds of thousands of years, but another to see the change happening in decades or centuries. Professor Jay Malcolm and others reported in *Habitats at Risk* that ecosystems are changing "ten times faster than the rapid changes during the recent postglacial period." Professor Anthony Barnosky wrote in *Heatstroke* that "earth has not experienced a similarly fast rate of climate change within at least the last 60 million years"—far longer than the ice ages, which only go back 2 or 3 million years. He added that the extinction rate of the past four hundred years, including the current global warming period, is 17 to 377 percent faster than normal for the

Yellow-bellied marmots thrive at high elevations, where habitat is now altered by the warming climate.

previous 65 million years of fossil record. Not only is the rate of change far faster now, but migrations and adaptations that plants and animals routinely made in the past are impossible in habitats fragmented by development, roads, fences, and farms. Biologists, such as Reed Noss and Allen Cooperrider, tell us that for the web of life to function and for species to survive, we must create protected corridors in order to reconnect those habitats, and increase the size of our reserves.

Altering the climate is the greatest experiment humanity has ever undertaken, and it's an unintentional experiment. Our ability to stop the avalanche of destructive consequences is limited. It's like being roped up together

Named for the famous glaciologist François Matthes, the Matthes Glacier is rapidly growing thinner and smaller. Here at Lobe Lakes, dwarf huckleberries, willows, and whitebark pines colonize lands once covered by the glacier in the ice ages.

and plodding toward hidden crevasses, with no opportunity for cautious, informed, or unwilling participants to choose a safer course for themselves and their children.

The most important point is this: how far we take the experiment is still up to us and to our political leaders. In the decisions we collectively now face, we can choose either self-restraint or aggrandizement, scientific analysis or ignorance. We can choose political governance that favors life, or that consigns the fate of the earth to the forces of vested economic power.

Perhaps all of this takes us a long way from the glaciers I've come to photograph, but now that I've seen them, I also see that those choices are as much a part of the ice as the rocks on its surface and the melting water inside.

Up there on Mount Maclure, with Greg Stock measuring the receding ice, and with a view across that great national park and toward all that lay below, I felt that the melting of the glaciers was telling us something we needed to know.

The North Palisade, center left, and Thunderbolt Peak, center right, collect deep snows in winter that nourish the North Palisade Glacier, which lies at the base of the cliffs. Fresh snow covers the slopes below in late May.

CHAPTER SIX

NORTH PALISADE

I saved the best until last. And I knew that now was the time to see and enjoy it.

Back in May my wife, Ann, and I had skied up Big Pine Creek to see the North Palisade Glacier in its formidable winter garb, but we were unable to actually reach the ice. To do that, September was the best time to go.

The snow had melted. The mosquitoes were done. The days warmed with the pungent smells of slightly rotting foliage along streams where the willows had turned yellow. Aspen leaves quivered one last time and dropped to the ground, golden but ready to return to the soil and nourish other lives in the years to come. The dawns were lacquered with frost that melted in the first luxuriant touch of sun, and the afternoons intoxicated me with a caress of warm air and with memories of all things autumn, bittersweet as it is. The impending chill and shortening days of fall signaled for me just how precious life is, because its time is short. In this way and others, the path of autumn now seemed to intersect with the path of glaciers in my life.

I was better prepared than I had been in the spring, when Ann and I set out with skis and heavy packs to get photos of the North Palisade: my load was now lighter, my path cleared of snow, my body acclimatized by two dozen high-country trips. But this was the big one. This largest glacier of the Sierra flows from its namesake peak, the range's third-highest at 14,242 feet. Unlike Mounts Whitney and Williamson, which have gentler profiles on some sides, the North Pal juts up like a monument for all to see, a challenge to mountaineers. It shadows an assemblage of peaks around it, each a paragon of mountain grandeur in its own right: Starlight, Thunderbolt, and Winchell to the northwest, Polemonium, Sill, Norman Clyde, and Middle Palisade to the east and south.

Not long ago, a suite of Palisades glaciers linked almost continuously for seven forbiddingly rugged miles along the Sierra crest, but no longer. Now the lineup is fragmented by melted-out ridgelines that taper down from the various peaks and split the shrinking array into orphaned cirque basins, each discrete, on its own, fading.

But for now, the North Palisade Glacier still spans the meta-cirque of the mountain's northeast flank, an impressive sight, from Sill over to Thunderbolt. The glacier remains about a half-mile wide and nearly a mile long, from the yawning bergschrund at its top to the iceberg-dotted lake at its bottom. Unlike most other Sierra glaciers or their remnants, the North Palisade still has the power of great mountain forces. It still has the feel of a glacier that shapes its landscape—of a creation story that's real and unfolding.

A long day's hike took me to the end of the main trail at Sam Mack Meadow, then up a steep path to the boulder rubble of a five-hundred-foot moraine, then on a rock-to-rock climb, and finally to a ridge overlooking the lower limits of the glacier. The lake lay below, gray-green with rock dust ground fine by the grating of boulders against boulders. At sunset I admired the view of the glacier's slow-motion flow from headwall to moraine and tried to imagine how long it would take for the ice at the bergschrund to creep all the way down to the lake and the terminal moraine. Concentric semicircles, each visible in a line of rock and stones, seemed to indicate the glacier's annual movement, looking from where I stood like the ridges of a giant clamshell.

Forest fires in Sequoia National Park had sent a haze over the crest, and the day's last sunlight angled in piercing rays through the faintly pungent smoke. Like the melting of the glaciers, the gauzy atmosphere was an indicator of the changing climate that's bringing more and larger wildfires to the superheated, dried-up forests below.

I pitched my tent on a ledge and lay down, tired to the bone at the end of an active day and also an active summer and fall. I was spared the usual howl of wind at that elevation; the night was utterly peaceful, not even a single mosquito whining outside. Yet somehow I lay sleepless with the intensity of my ledgy perch and of feelings that welled up about this place, and about this chapter in its history. I didn't want to go to sleep. I wanted the moment, the hour, the night, and the next day to last longer than I knew they would.

With the sunrise I first had to climb down the side of the moraine to the surface of the lower glacier on a route riddled with hidden hazards. Left in unconsolidated piles by the glacier, the rocks and boulders were intrinsically unstable. Any one of my steps could trigger a tip, or the tumbling of a granite behemoth, or a crushing slide. I chose my route carefully, making sure I didn't step where a rock could drop on top of me, and I avoided handholds that might pull a boulder down.

I aimed for the base of the moraine on the near side of the lake, where I planned to take a picture, but as luck would have it the incongruities of my route led me farther toward the body of the glacier above the lake. When I reached the ice edge I hoisted myself up on a big old erratic in order to eyeball a truer view. What I had thought was simply the moraine's boulder field slanting into the lake—where I had been headed—was in fact a boulder garden entirely underlain by ice dozens of feet thick—an "ice-cored" moraine. What I had thought was rock was in fact glacier. Looking back there, I saw that the hidden ice was calving bit by bit into the lake, in the

The North Palisade Glacier begins at the nearly vertical walls of the mountain and ends in the glacial margin lake below, shown here in September. The rock moraine on the near side of the lake is underlain by ice. Smoke from forest fires on the west side of the mountains hazes the sky.

A stream of autumn meltwater has refrozen overnight on the lower end of the North Palisade Glacier.

This large table rock fell after the ice column that supported it for many years melted. In the background, Mount Sill rises in morning sunlight above the glacier.

Carrying rocks and boulders, the North Palisade Glacier sweeps down through its extensive cirque basin.

process shedding its thick skin of rocks and boulders into the murky, frigid depths.

Thankful for the way my route had taken me, I strapped on my crampons and started up the ice, admiring its gelid features: diminutive rivers of crystal clear meltwater that had refrozen overnight, mosaics of rock buried in glaze, table rocks perched on platforms of ice. Some of these rocks measured ten feet wide on pedestals half that size. This unlikely placement of the large and heavy atop the small and light is an artifact of rocks' riding the glacier's slow flow down the mountain, all the while shading and insulating the ice beneath them. While the sun-warmed ice around the rocks melted, the ice beneath them remained solid. In effect, the height of each pedestal indicated the amount the glacier had thinned during the period that the rock was in place and across the distance the rock had come. In time, each table rock was destined to fall from its impossible perch. As I watched, one three-foot boulder tipped in the melting morning sun and smashed onto the ice immediately below. The ice surrounding it might melt to form another pedestal as the boulder continued its journey, eventually to be dumped into the lake by the expiring flow of ice.

The slope of the glacier steepened, so I clutched my axe more diligently and weighed the prospects, should I fall, for self-arrest. Given the metallic hardness of the morning ice, stopping my slide would be no cinch. But I continued anyway, because the runout zone below didn't look too bad. All the while the landmarks below grew smaller while the headwall above grew larger, its mammoth dripping icicles like a gargantuan pipe organ ready to blast out a hymn.

I stopped to rest. But the slope angled sharply, and my crampons barely gripped the arctic surface, leaving me with tense and tiring legs. After one more pitch full of heavy breathing, I arrived at the top of the glacier and at the ice-mounded lip of the bergschrund—a ten-foot-wide gap between the glacier and the mountain's headwall.

Dropping to depths of ten and twenty feet, and then continuing with Swiss-cheese cavities and black, bottomless holes that had melted, traplike, into the ice, this impressive feature brought me to a halt. The rocks I had negotiated near the lake were unstable because they had been deposited as a moraine only ten thousand years ago; rocks in the bergschrund might have been deposited only a day ago, and they promised to settle further at the slightest provocation.

I stared down into the eerie blend of brilliance and gloom. Then I stared up at the polished, icy surface of a couloir—a narrow, iced flume that veered sharply toward the summit's U-notch, famous as a climber's route to the top.

I turned and looked back down the North Palisade Glacier, and I thought about the years required for the ice to move from top to bottom. Given today's rate of warming, it's doubtful that the ice there at the bergschrund will

The bergschrund of the North Palisade separates the main ice mass from the couloir of the U-notch that rises toward the summit of the Sierra Nevada's third-highest peak.

Rising above the North Palisade Glacier, the North Palisade and Thunderbolt Peak catch the morning's first light.

last long enough to endure the glacier's long downhill slide to the lake. The glacier will likely be gone before the ice where I stood can get to the bottom.

Whatever will take the North Palisade Glacier's place will be beautiful as well: the fell-fields and tumult of rock, the meadows evolving with a pileup of windblown dust, the streams that may gradually surface when the underground cavities fill with sediment. Even the shape of the mountains will be altered in the endless cycles of change. But I was grateful to be there in the autumn of 2010, when I could still stand on the ice and snow of a glacier.

Before dawn, a full moon sets over the Palisade Crest and its glacier, which are both freshened with new snowfall.

ABOUT THE PHOTOGRAPHS

All photos were taken with Canon A-1 cameras and Fuji Velvia film, which captures natural scenes with great accuracy and authenticity. I used fixed focal length lenses from 17 to 100mm and shot most photos from a lightweight tripod, using time exposures to record some scenes in low light. No photos were altered in color or composition after they were shot, no artificial light was used, and no double or combined exposures were taken. For printing, the publisher scanned the slide transparencies to match the color and content of the slides themselves.

Because the Sierra glaciers are nearly all on north- and east-facing slopes within cirque basins near the tops of mountains, the most exquisite light shines on them at sunrise, when the rays are low and golden as they burn through the lower atmosphere from the east. At that moment the shadows are deep, the shading dark, the contrast brilliantly edged in the first direct light. To capture it on film meant that I needed to be high on the mountain, if not on the glacier itself, at that magic moment when the first rays struck the peaks. So I camped as high as possible, given the need for a nominal place to sleep in rugged terrain. In what became a pattern, I awoke an hour before dawn, gathered my camera kit and my headlamp and began my sprint upward; from first light to direct sunlight I had fifty minutes. Pictures were taken at other times of the day, but most were shot early in the morning.

SOURCES

Badè, William Frederic. *The Life and Letters of John Muir*. Boston: Houghton Mifflin, 1924.

Barnosky, Anthony D. *Heatstroke: Nature in an Age of Global Warming*. Washington, D.C.: Island Press, 2009.

Basagic, Hassan Jules IV. "Quantifying Twentieth Century Glacier Change in the Sierra Nevada, California." Master's Thesis, Portland State University Department of Geography, 2008.

Basagic, Hassan J., and Andrew G. Fountain. "Quantifying 20th Century Glacier Change in the Sierra Nevada, California." *Arctic, Antarctic, and Alpine Research* 43, no. 3, 317–330.

Broder, John M. "Climate Change Seen as Threat to U.S. Security." *New York Times*, August 8, 2009.

California Climate Change Center. "Our Changing Climate: Assessing the Risks to California." Sacramento: California Energy Commission, 2006.

California Department of Water Resources. *California Water Plan Update*. Sacramento: the Department, 2005.

Guyton, Bill. *Glaciers of California*. Berkeley: University of California Press, 1998.

Hayhoe, Katharine, et al. "Emissions, Pathways, Climate Change, and Impacts on California." *Proceedings of the National Academy of Sciences*, August 24, 2004.

Hill, Mary. *Geology of the Sierra Nevada*. Berkeley: University of California Press, 1975.

Howat, Ian M., et al. "A Precipitation-dominated, Mid-latitude Glacier System: Mount Shasta, California." *Climate Dynamics* 28, No. 1, 85–98 (2005).

Keyes, Scott. "Report: Meet the 2010 GOP Freshman Class." *Think Progress*, http://thinkprogress.org/2010/11/03/gop-frosh-class, Nov. 11, 2010.

Malcolm, Jay R. et al. *Habitats at Risk*. Gland, Switzerland: World Wide Fund for Nature, 2002.

Millar, Constance, David Clow, Jessica Lundquist, and Robert Westfall. "Rock Glaciers and Periglacial Rock-Ice Features in the Eastern Sierra Nevada, California," http://www.fs.fed.us/psw/publications/millar/posters/millar_etal_poster_paclim2006.pdf (accessed June 17, 2011). USDA Forest Service Sierra Nevada Research Center.

Muir, John. *The Mountains of California*. 1894. Reprint, New York: Doubleday, 1961.

Price, Larry W. *Mountains and Man: A Study of Process and Environment*. Berkeley: University of California Press, 1981.

Rosenthal, Elisabeth. "For Many Species, No Escape as Temperature Rises." *New York Times*, Jan. 21, 2011.

———. "Imperiled Birds on a Warming Planet." *New York Times*, January 26, 2011.

Schweiger, Larry J. *Last Chance: Preserving Life on Earth*. Golden, CO: Fulcrum, 2009.

Secor, R. J. *The High Sierra: Peaks, Passes, and Trails*. Seattle: The Mountaineers, 2009.

Selters, Andy, and Michael Zanger. *The Mt. Shasta Book*. Berkeley: Wilderness Press, 2006.

Sharp, Robert P. *Living Ice: Understanding Glaciers and Glaciation*. Cambridge: Cambridge University Press, 1988.

Sheats, Paul. "John Muir's Glacial Gospel." *The Pacific Historian* 29 (Summer/Fall 1985).

Sierra Nevada Alliance. *Sierra Climate Change Toolkit*. South Lake Tahoe: the Alliance, 2005.

United Nations Intergovernmental Panel on Climate Change. *Climate Change 2007*, http://www.ipcc.ch/publications_and_data/ar4/syr/en/contents.html (accessed June 17, 2011). Geneva, Switzerland: the Panel, 2007.

Young, Samantha. "Glaciers on California's Mt. Shasta Keep Growing." *USA Today*, July 8, 2008.

ACKNOWLEDGMENTS

Many thanks to my wife, Ann Vileisis, who accompanied me on some of the glacier trips, including the hardest one. Unfortunately, other important commitments and a challenging book of her own prevented her from going on all the outings. But Ann graced me with valuable ideas throughout, edited the manuscript, and supported me in every way, whether she was with me or not.

I cannot offer enough thanks, respect, and admiration to publisher Malcolm Margolin of Heyday, who saw the promise in this book idea before I did. Much-deserved thanks also go to editors Gayle Wattawa and Jeannine Gendar, production director Diane Lee, designer Lorraine Rath, marketing director Natalie Mulford, and indeed all the Heyday staff. The role of Sierra College Press in supporting the book is much appreciated. Nat Hart read the manuscript and contributed his sensibilities as a veteran English professor, fine writer, and good friend.

Yvon Chouinard and his company, Patagonia, helped with essential financial support, as did Norton, Adam, and Melanie Smith of the Whole Systems Foundation; Norton also joined me for my hike to the Goethe and Matthes Glaciers and three days of thoughtful conversation and friendship. Paul and Catherine Armington of the Evenor Armington Fund likewise contributed to my work. Greg Knight and Marieta Staneva of the Knight-Staneva Foundation for Sustainability and Future Environments supported my outreach efforts with the book.

Photographer Jeff Pflueger and Outward Bound professional Nettie Pardue joined me on my crucial Mount Shasta trip and shared their knowledge during a weekend of exceptional companionship. Geographer Hassan Basagic's superb thesis about Sierra glaciers was informative in many ways, and he graciously read and reviewed my entire manuscript. The historic glacier photos that I used were gleaned from the collection that Hassan amassed while doing his research through Portland State University. Glaciologist Slawek Tulaczyk checked portions of the manuscript. Greg Stock of

Yosemite National Park generously invited me to join him on a research expedition to the Lyell and Maclure Glaciers, and he pointed the way to the Maclure ice cave; Josh Helling and Steve Bumgardner were great company on that trip as well. Greg later read the manuscript with a scientist's eye for accuracy. Important tips about Mount Shasta were provided by Jenn Carr of Shasta Mountain Guides, Leif Voeltz of The Fifth Season in Mount Shasta City, and Neil Woodruff of Sierra Wilderness Seminars. James Wilson and Kent Barton of Wilson's Eastside Sports shared their vast expertise about the southeastern Sierra. Steve Sweringen of the Camera Clinic in Sparks kept my photo gear in perfect shape. Many thanks also to the authors of all the fine works cited on my sources pages.

Index

ABOUT THE AUTHOR AND PHOTOGRAPHER

Tim Palmer has written twenty-two books about the American landscape, rivers, conservation, and adventure travel. He has spent forty years hiking, exploring, and photographing in California and the rest of America, and his lifelong passion has been to write and speak on behalf of conservation.

Tim's recent books include *Rivers of California: Nature's Lifelines in the Golden State* (Heyday, 2010) and *Luminous Mountains: The Sierra Nevada of California* (Heyday, 2008). His *Field Guide to California Rivers* was published by the University of California Press in 2012.

Tim's *California Wild* won the Benjamin Franklin Award for nature books in 2005. *The Heart of America: Our Landscape, Our Future* won the Independent Publisher Book Award for travel/essay books in 2000. Other awards include the National Outdoor Book Award (*The Columbia*, 1998) and the Director's Award from the National Park Service (*Yosemite: the Promise of Wildness*, 1997).

Recognizing his accumulated contributions in writing and photography, American Rivers gave Tim its first Lifetime Achievement Award, in 1988, and California's Friends of the River has recognized him with both its highest honors, the Peter Behr Award in 2002 and the Mark Dubois Award in 2010. *Paddler* magazine named him one of the ten greatest river conservationists of our time and in 2000 included him in the "100 greatest paddlers of the century." In 2005 Tim received the Distinguished Alumni Award from the College of Arts and Architecture at Pennsylvania State University. Topping these honors, Tim received the National Wildlife Federation's Conservation Achievement Award ("Connie") for communications in 2011.

In addition to freelance work, Tim does contract writing, photography, and river studies for conservation organizations. He frequently speaks and gives slide shows for universities, conservation groups, outdoor clubs, workshops, and conferences nationwide. You can learn more about his work at www.timpalmer.org.

ALSO BY TIM PALMER

Rivers of California

Luminous Mountains: The Sierra Nevada of California

Field Guide to California Rivers

Trees and Forests of America

Rivers of America

California Wild

Oregon: Preserving the Spirit and Beauty of Our Land

Pacific High: Adventures in the Coast Ranges from Baja to Alaska

Lifelines: The Case for River Conservation

Endangered Rivers and the Conservation Movement

The Heart of America: Our Landscape, Our Future

America by Rivers

The Columbia

Yosemite: The Promise of Wildness

The Wild and Scenic Rivers of America

California's Threatened Environment

The Snake River: Window to the West

The Sierra Nevada: A Mountain Journey

Youghiogheny: Appalachian River

Stanislaus: The Struggle for a River

Rivers of Pennsylvania

SIERRA COLLEGE PRESS

In 2002, the Sierra College Press was formed to publish *Standing Guard: Telling Our Stories* as part of the Standing Guard Project's examination of Japanese American internment during World War II. Since then Sierra College Press has grown into the first complete academic press operated by a community college in the United States.

The mission of the Sierra College Press is to inform and inspire scholars, students, and general readers by disseminating ideas, knowledge, and academic scholarship of value concerning the Sierra Nevada region. The Sierra College Press endeavors to reach beyond the library, laboratory, and classroom to promote and examine this unique geography.

For more information, please visit www.sierracollege.edu/press.

Editor-in-Chief: Gary Noy

Board of Directors: Bright Rope, Rebecca Bocchicchio, Julie Bruno, Keely Carroll, Kerrie Cassidy, Charles Dailey, Mandy Davies, Frank DeCourten, Daniel DeFoe, Danielle DeFoe, Tom Fillebrown, Brian Haley, Robert Hanna, Rick Heide, Carol Hoge, Jay Hester, Roger Lokey, Joe Medeiros, Lynn Medeiros, Sue Michaels, Mike Price, Jennifer Skillen, Randy Snook, Barbara Vineyard

Advisory Board: Terry Beers, David Beesley, Patrick Ettinger, Janice Forbes, Tom Killion, Tom Knudson, Gary Kurutz, Scott Lankford, John Muir Laws, Beverly Lewis, Malcolm Margolin, Mark McLaughlin, jesikah maria ross, Michael Sanford, Lee Stetson, Catherine Stifter, Rene Yung

Special thanks to our major financial supporters: Sierra College Friends of the Library and the Rocklin Historical Society

HEYDAY
into California

About Heyday

Heyday is an independent, nonprofit publisher and unique cultural institution. We promote widespread awareness and celebration of California's many cultures, landscapes, and boundary-breaking ideas. Through our well-crafted books, public events, and innovative outreach programs we are building a vibrant community of readers, writers, and thinkers.

Thank You

It takes the collective effort of many to create a thriving literary culture. We are thankful to all the thoughtful people we have the privilege to engage with. Cheers to our writers, artists, editors, storytellers, designers, printers, bookstores, critics, cultural organizations, readers, and book lovers everywhere!

We are especially grateful for the generous funding we've received for our publications and programs during the past year from foundations and hundreds of individual donors. Major supporters include:

Anonymous; Evenor Armington Fund; James Baechle; Bay Tree Fund; S. D. Bechtel, Jr. Foundation; Barbara Jean and Fred Berensmeier; Berkeley Civic Arts Program and Civic Arts Commission; Joan Berman; Peter and Mimi Buckley; Lewis and Sheana Butler; California Council for the Humanities; California Indian Heritage Center Foundation; California State Library; Keith Campbell Foundation; Candelaria Foundation; John and Nancy Cassidy Family Foundation, through Silicon Valley Community Foundation; Center for California Studies; Compton Foundation; Nik Dehejia; Frances Dinkelspiel and Gary Wayne; George and Kathleen Diskant; Donald and Janice Elliott, in honor of David Elliott, through Silicon Valley Community Foundation; Euclid Fund at the East Bay Community Foundation; Eustace-Kwan Charitable Fund; Federated Indians of Graton Rancheria; Mark and Tracy Ferron; Judith Flanders; Furthur Foundation; The Fred Gellert Family Foundation; Wallace Alexander Gerbode Foundation; Wanda Lee Graves and Stephen Duscha; Alice Guild; Coke and James Hallowell; Carla Hills; Sandra and Charles Hobson; G. Scott Hong Charitable Trust; James Irvine Foundation; Kendeda Fund; Marty and Pamela Krasney; Guy Lampard and Suzanne Badenhoop; LEF Foundation; Judy McAfee; Michael McCone; Joyce Milligan; National Endowment for the Arts; National Park Service; Steven Nightingale; Theresa Park; Patagonia, Inc.; Pease Family Fund, in honor of Bruce Kelley; The Philanthropic Collaborative; PhotoWings; Alan Rosenus; The San Francisco Foundation; San Manuel Band of Mission Indians; Savory Thymes; Hans Schoepflin; Contee and Maggie Seely; Sandy Shapero; William Somerville; Martha Stanley; Stanley Smith Horticultural Trust; Stone Soup Fresno; Roselyn Chroman Swig; James B. Swinerton; Swinerton Family Fund; Thendara Foundation; Tides Foundation; Lisa Van Cleef and Mark Gunson; Marion Weber; Whole Systems Foundation; John Wiley and Sons; Peter Booth Wiley and Valerie Barth; Dean Witter Foundation; and Yocha Dehe Wintun Nation.

Getting Involved

To learn more about our publications, events, membership club, and other ways you can participate, please visit www.heydaybooks.com.